Blockchain:

Bitcoin, Ethereum, Smart Contracts, Cryptocurrencies and Everything about the Fintech Explained

Shane Bock

© 2017

☐ © Copyright 2017 by All rights reserved.

This document is geared towards providing exact and reliable information in regards to the topic and issue covered. The publication is sold on the idea that the publisher is not required to render accounting, officially permitted, or otherwise, qualified services. If advice is necessary, legal or professional, a practiced individual in the profession should be ordered.

From a Declaration of Principles which was accepted and approved equally by a Committee of the American Bar Association and a Committee of Publishers and Associations.

In no way is it legal to reproduce, duplicate, or transmit any part of this document by either electronic means or in printed format. Recording of this publication is strictly prohibited and any storage of this document is not allowed unless with written permission from the publisher. All rights reserved.

The information provided herein is stated to be truthful and consistent, in that any liability, regarding inattention or otherwise, by any usage or abuse of any policies, processes, or directions contained within is the solitary and utter responsibility of the recipient reader. Under no circumstances will any legal responsibility or blame be held against the publisher for any reparation, damages, or monetary loss due to the information herein, either directly or indirectly.

Respective authors own all copyrights not held by the publisher.

The information herein is offered for informational purposes solely and is universal as so. The presentation of the information is without a contract or any guarantee assurance.

The trademarks that are used are without any consent, and the publication of the trademark is without permission or backing by the trademark owner. All trademarks and brands within this book are for clarifying purposes only and are the owned by the owners themselves, not affiliated with this document.

Disclaimer and Terms of Use: The Author and Publisher have strived to be as accurate and complete as possible in the creation of this book, notwithstanding the fact that he does not warrant or represent at any time that the contents within are accurate due to the rapidly changing nature of the Internet. While all attempts have been made to verify information provided in this publication, the Author and Publisher assumes no responsibility for errors, omissions, or contrary interpretation of the subject matter herein. Any perceived slights of specific persons, peoples, or organizations are unintentional. In practical advice books, like anything else in life, there are no guarantees of results. Readers are cautioned to rely on their judgment about their individual circumstances and act accordingly. This book is not intended for use as a source of legal, medical, business, accounting or financial advice. All readers are advised to seek services of competent professionals in the legal, medical, business, accounting, and finance fields.

Contents

Introduction 6

Chapter 1: What is Blockchain? 9

Chapter 2: How does Blockchain work? 18

Chapter 3: Bitcoin and the Rest of the Cryptocurrencies . 28

Chapter 4: Limitations of Blockchains 45

Chapter 5: Applications... 54

Conclusion... 61

Author's Notes 65

Introduction

In recent years, no other technology has been the subject of such fervent debate. As much as some might want to disagree, the rise in popularity of cryptocurrencies cannot be ignored. In the world today, there are a number of billion dollars businesses out there that do accept Bitcoin as a form of payment. These include Dell, Reddit, Expedia, PayPal, and most recently, even Microsoft.

As important as they may be, Bitcoin and any other cryptocurrencies cannot be where they are now if not for the underlying technology known as blockchain. I am sure it is not that hard to see what the internet has done to the media. How we receive information has been changed significantly from how it used to be. The internet has since become a major influence on our lives, from how we buy goods and services, to the ways we socialize with friends.

Now, visualize this.

The blockchain technology will do the same to the financial system as what the internet did to media

Things will never be the same again. The blockchain will be the next hot topic to be widely discussed about. In fact, it is hailed as the next coming of the internet. As important as it is, the blockchain might not be a very familiar term to the general population but there should not be any doubts that most people have heard about the Bitcoin.

Well, the blockchain is basically the underlying technology that makes Bitcoin. Looking at how much the Bitcoin is worth now; we can tell how important the blockchain is or at least, is going to be.

Blockchain is growing in significance as it gets more recognized for its functions. It is steadily transforming the financial relationships between people and businesses globally. Increasingly, organizations have to explore what this revolutionary technology will mean for their business or risk having their business disrupted by this technology.

In 2016, a record-breaking of USD$1 billion has been pumped into Bitcoin and blockchain-related tech start-ups. The banks are rushing to incorporate this technology into their banking system which is believed to be able to help them save billions of dollars annually. Four of the biggest global banks, led by Swiss bank UBS, have developed a "Utility Settlement Coin" (USC), which is the digital counterpart of each of the major currencies backed by central banks. The Australia Post has released plans for developing a blockchain-based e-voting system for the state of Victoria.

As we can see, the possibilities of the blockchain are enormous and it seems that almost any industry that deals with some sort of transaction, which would likely mean any industry, can and will be disrupted by the blockchain revolution. As a result, it is likely that many of the 3rd party intermediary industries will face job losses since intermediaries will be needed a lot less.

From the above, we can see that some emphasis has been placed into blockchain related activities but honestly, will the blockchain really be the next big thing or is it just a passing hype?

Some of us might still remember the Dot-com bubble in 1995. Everyone thought it was real and going to stay but it busted. Does the blockchain carry the same fate as the Dot-com era? Are the blockchain related companies too excited that they are blinded to realise that the blockchain is probably just a passing fad? Or are these companies just trying to make a quick buck before the bubble pops?

No one knows for sure but looking at the potential of the blockchain, it seems that the world holds high expectations.

Chapter 1: What is Blockchain?

A blockchain to put it very simply, is a digital ledger. It is a continuously growing list of records which are linked and secured using cryptography. The records are termed as blocks and each block is linked to a previous block, hence the name blockchain. By design, blockchains are inherently resistant to modification of the data.

Functionally, a blockchain can be used for recording transactions between two parties efficiently and in a verifiable and permanent way. It is typically managed by a peer-to-peer network which can be anyone that is willing to take up the job of validating new blocks. These people are given the names of miners. Once recorded, the data in any given block are sent to miners in the network. Every miner holds the same data and any attempt to alter it can be easily discovered simply by comparing the data with other miners in the network. As you might know already, the data is not controlled by a centralised organisation like banks but distributed out to interested people like you and me. This gives a lot of transparency and clarity to the transaction made.

Decentralized systems

The greatest characteristic of a blockchain is basically its decentralised network. Decentralized technology allows us to store assets in a distributed network that can be accessed over the Internet. The assets can be anything from cryptocurrencies like Bitcoin to a simple piece of will. Through decentralized technology, the owner has direct control over their asset through their private key, which is directly linked to the asset. The owner can transfer the asset to anyone whenever desired.

The blockchain technology has proven itself to have the potential to bring revolutionary changes in the existing industries like banking and finance. As this technology lays ground in these fields the individuals will have the direct control of their assets, then they wouldn't need a third party middle man which is taking a huge fee for the transaction charges. The individual can directly indulge in the transaction giving minimal fee. Banks have already acknowledged this upcoming crisis and are working to embrace this disruptive technology.

The interesting point about a decentralised system is about not placing your trust in a single entity like a government, a bank, or a company. The trust is actually not required anymore due to the characteristic of decentralisation. There are many more points to a decentralised system which we will discuss as follows.

(a) Empowered users:

Users keep control of all their information and transactions and This is only possible with a decentralised system. A centralised system will allow the company or bank to do the opposite since it is only one entity keeping your information.

(b) Fault tolerance:

Decentralized systems are less likely to fail accidentally because they rely on many separate components that works independently rather than as one entity. The end results are actually based on the result that everyone derives upon rather than the single result derived by an individual person, bank or government.

(c) Durability and attack resistance:

Since blockchain does not have a central point of control, it is able to survive a malicious attack. The decentralized systems are virtually almost impossible to attack and manipulate. It would probably take too much effort or resources to bring it down even if it is theoretical possible.

(d) Free from scams:

It is much harder for scammers in decentralized systems to work in a way that will benefit them since everyone on the network keeps the same ledger. Any difference can be quickly and easily spotted without any need for arguments.

(e) Removing third-party intervention:

This technology enables users to make an exchange without the intervention of a third party, thus eliminating all the risks involved like high costs, frauds or even human errors. This is almost as close to a real world transaction as it can get.

(f) Lower transaction costs:

By eliminating third-party intermediaries and overhead costs for transactions, blockchains have the potential to greatly reduce transaction fees

(g) Higher transaction rate:

Conventionally, it takes days for the third-party banks to process the transaction. Blockchain transactions can reduce transaction times to as low as minutes and they can be processed anytime.

(h) Transparency:

Changes made to public blockchains are publicly viewable by all parties within the network thereby creating transparency. All transactions are cannot be altered or deleted.

(i) Authenticity:

Because of the decentralized system, the blockchain data can be complete, consistent, accurate and readily available.

Although simple, we see many benefits to having a decentralised network. The system allows individuals to hold power rather than a single entity like bank or government which is probably, what a free world and power to the people is really all about.

Who Invented Blockchain and some Background

The first work on a cryptographically secured chain of blocks was described around 1991 by Stuart Haber and W. Scott Stornetta. Many years later, an anonymous person or group known by the Japanese name of Satoshi Nakamoto, published blockchain's original definition in 2008. Satoshi is also famous for being the creator of Bitcoin. Using blockchain technology, Bitcoin is the first digital currency to solve the double spending problem, without the use of a trusted authority or central server.

Up until now, the world believed that Satoshi is the original creator of the blockchain technology. But as it turns out, this might not necessarily be the case. An Inventor by the name of Dr. Kelce Wilson claimed that he first invented blockchain back in 2000 and originally patented his idea 6 months before Satoshi Nakamoto. However, Satoshi remains as the more well-known creator of blockchain probably due to the popularity of Bitcoin.

The words block and chain were used separately in Satoshi Nakamoto's original paper when it was published, before becoming a single word, blockchain, by 2016. The blockchain technology has gained much attention in recent years and its file size has increased significantly. In August 2014, the bitcoin blockchain file size reached 20 gigabytes and this number jumped to almost 30 gigabytes by January 2015. As of date, January 2016 to January 2017, the bitcoin blockchain grew from 50 gigabytes to 98 gigabytes in size.

Problem that Blockchain solved

We have mentioned a great deal about blockchain technology but what is really so great about this technology? What exactly is the problem that blockchain solve which the current technology today cannot properly address?

The most common or well-known use for blockchain right now is digital currency. Bitcoin without saying will be the best example. However, we already have currencies and foreign exchanges in place so why would we still want digital currencies? This comes down to the fact that today's currencies are centralized and controlled. For centralized banks to handle the money, there are generally two main issues.

Cost

International money transfer can be expensive and slow. Though we might not see it, there might be more than a bank or organization involved thus increasing the cost and time taken to make the transfer.

Trust

There are not many choices other than to trust the banks. As individuals, we do not really have much choice unless we physically make the transaction with the other party. Realistically, it is possible to misplace all your money. You do not have any control over this. Since the money is not directly given to the receiver, you have to trust a middleman when wiring your money around.

Blockchain technology solves these problems really well with the use of cryptography. For instance, anyone can create a Bitcoin wallet on their smart phone at zero cost to start receiving and spending money, with low transaction costs (a few cents today). Trust is not required in this transaction. It is just like paying over a counter with real hard cash. The money is given along with receiving the goods.

With blockchain, there is total transparency. If the money you donated online to a certain organization in another part of the world was traceable, it would be much harder to cheat the people in need. If any money went missing, everyone would know about it, and the organization could be held accountable. Another example could be a microfinancing institution. It would be good to know that the money you loan out is going to the correct people. Corruption and fraud could be made a lot harder to be committed with blockchain since every transaction is recorded into the block and linked to the previous.

The applicability of blockchains may include everywhere that many people may want to interact with a computer database. It is easy to imagine a tremendous breadth and depth of potential applications and markets especially when data tracking is heavily involved.

Chapter 2: How does Blockchain work?

There are basically three principal technologies that make up the blockchain.

1) A distributed network with a shared ledger

2) Private Key cryptography

3) An incentive to service the network's transactions, record-keeping and security.

None of them are new. However, it is their combination and application that is something new. The following is an explanation of how these technologies orchestrate together to make blockchain so amazing.

Cryptographic keys

When two parties make a transaction over the blockchain, they are actually sending them to a Public Key which is in the form of a hashed version. There is another key which is hidden from them, which is called the Private Key. The purpose of the Private Key is to derive the Public Key. The private key is private as the name suggests and only the person holding it will know it. You can know your own Private Key and everyone else on the blockchain knows their own Private Key.

Both the Private Key and the Public Key are usually represented using a format consisting of letters and numbers known as the Wallet Import Format (WIF).

A sample of Private Key in WIF will look like this

5AkeCFO9rMksESpoPoS5BGks4LkDefQsf3dX1dgaRAfbxEGhyRS

The Private Key is used to generate a signature for each blockchain transaction a sender sends out. In short, you sign the cryptocurrencies you send to others using a Private Key. Unless others have your Private Key, else it is almost impossible to hack the code and make transactions in your place.

Identity

Although the Private and Public Keys ensure a strong control of ownership, it is not enough to secure digital relationships. While authentication is solved, it must be combined with a means of approving transactions.

For blockchains, this begins with a distributed network. The benefit and need for a distributed network ensures that everyone in the network stores a record of the transaction blockchains.

If a system gets hacked and data altered, it is unlikely the others in the network will get affected since the network is distributed and decentralised. It takes too much effort and resources for the hacker to alter the entire network.

In short, the size of the network is important to secure the network.

That is one of the most attractive characteristics of the blockchain. It is so ridiculously huge and requires so much computing power. When cryptographic keys are combined with this network, an impenetrable form of digital security emerges.

The Incentive

Let's be realistic about this. No one will spend their time on the network to simply spend computing power to solve mathematical problems for the good of mankind unless there are some forms of rewards or benefits to them while doing this. In another words, how do you attract computing power to service the network to make it secure?

Mining

For open, public blockchains, this involves something called mining. The people or computing powers that carry out the mining are called miners. Just like real miners, they are given the opportunity to be rewarded with finding "gold".

With blockchains, by offering your computer processing power to service the network, there is a reward available for one of the computers. A person's self-interest is being used to help service the public need.

This is how bitcoin seeks to act as gold. Bitcoins must be unique to be owned and provide value. To achieve this, the nodes serving the network create and maintain a history of transactions for each bitcoin by working to solve mathematical problems.

The miners basically vote with their CPU power, expressing their agreement about new blocks or rejecting invalid blocks. When a majority of the miners arrive at the same solution, they add a new block to the chain. This block is time stamped, and can also contain data or messages.

Example of transaction using Bitcoin.

Let's say Jane owns some bitcoins. In order to transfer the coins to another entity, Jane will use the cryptographic keys to produce a digital signature on the statement to make a transaction.

Now, suppose Jane signs a statement on her own computer saying she wants to transfer some coins to Bobby but never did send the statement to Bob. In this case, clearly the coins have not been transferred out of her account. Therefore, this will not be witnessed by anyone on the distributed network and cannot be considered as a legitimate transaction.

Moving on, let's say Jane did send the signed statement to Bobby. However, this will not be considered transacted if Jane had signed another transaction statement saying that she wants to transfer the coins to Caroline, which she only sends to Caroline. If Bobby and Caroline both accepted these statements indicating that they have received the coins from Jane, then Jane will have effectively spent her coins twice!

Now, this is where the good point of a distributed global ledger comes in. If Jane wants to transfer her coins to Bobby, she must publish her statement authorizing the transfer to the blockchain. The miners who maintain the blockchain will only authorise this transaction if Jane has not yet transferred the coins to anybody else. Once Bobby sees the transaction appear in the blockchain, he can be confident that he is the new owner. Even if Jane later tries to sneakily produce a statement saying she transferred the coins to Caroline, it will never be approved into the blockchain because the transaction transferring to Bobby was published first.

How is mining done?

When a block of transactions is created, miners using computers put it through a process. They take the information in the block, and apply a mathematical formula to it, turning it into something else. That something else seems like a random sequence of letters and numbers commonly known as a hash. This hash is stored together with the block, at the end of the blockchain at that point in time.

Hashes have rather interesting properties. It is not that difficult to produce a hash from a collection of data like a bitcoin block, but it's basically impossible to figure out what the data was just by staring at the hash. And while it is fairly easy to produce a hash from a large amount of data, each hash is actually unique. If even just one character in a block is changed, its hash will change completely.

The transactions in a block are not the only data used by miners to generate a hash. Some other pieces of data are required to make the hash. One of the pieces of data is actually another hash, which is the hash of the last block stored in the blockchain.

Due to the fact that each block's hash is produced using the hash of the block before it, the hash acts like a digital version of a wax seal. It confirms that this block – and every block after it – is legitimate, because if it is ever tampered with, everyone would know.

If a hacker ever tried to fake a transaction by changing a block that had already been stored in the blockchain, that block's hash would change totally. If a miner checked the block's authenticity by running the hashing function on it, they'd find that the hash was of a different one from the one already stored along with that block in the blockchain. The block would immediately be spotted as a fake.

Because each block's hash is used to help produce the hash of the next block in the chain, tampering with a block would also make the subsequent block's hash wrong too. That would continue all the way down the chain, throwing everything out of track like the dominoes effect.

Competing for coins

The miners generally use software written specifically to mine blocks. In the case of Bitcoin, every time a miner successfully creates a hash, they are rewarded with 25 pieces of bitcoins, the blockchain gets updated, and everyone on the network will know about it.

Now, the problem is that it is relatively easy to produce a hash simply from a collection of data. With humans, it takes a long time if ever, but computers are really excellent at this. Therefore, the bitcoin network has to make it more challenging, or else everyone would be hashing hundreds of transaction blocks every second, and all of the bitcoins would be mined in minutes. The bitcoin protocol deliberately makes it more difficult, by introducing something called 'proof of work'.

The bitcoin protocol will not simply just accept any old hash. It requires that a block's hash has to look in a certain format. For instance, it must have a certain number of zeroes at the start. There is no way of telling what a hash might look like before it is produced, and as soon as you include a new piece of data in the mix, the hash will change totally to a different one.

The miners' job is not to meddle with the transaction data in a block, but they need to change the data they're using to create a different hash. They do this by using another random piece of data known as the nonce. This is used with the transaction data to create a hash. If the hash doesn't fit the required format, the nonce will be changed, and the whole thing is hashed again. It can take multiple attempts to find a nonce that actually work, and all the miners in the network are trying to do this thing all at the same time. Basically the miner with the best computing power will have the highest chance to find the correct hash and it is very common for miners to pool their computing power together to find that hash.

How long does it take to build a block?

One hour is the common answer to building a community standard of six blocks but this is not all to the story. Because blocks are discovered by a random process, there is no way of telling exactly how long it will take for 6 blocks to be chained. On average, it takes about 10 minutes to find each block. Do note that the difficulty of mining will increase if the blocks are increasing in size. However, with the enhancement of technology or computing power, it is definitely hard to tell if mining speed will increase or decrease.

Chapter 3: Bitcoin and the Rest of the Cryptocurrencies

What is Bitcoin?

We are probably more familiar with Bitcoin more than blockchain or other cryptocurrencies. But still, many folks are still quite confused about the difference between blockchain and Bitcoin. Some even has this idea that Bitcoin is blockchain. Honestly it is not easy to understand Bitcoin and blockchain because they are something new and complex. However, it is not that hard to differentiate between blockchain and Bitcoin. Simply put, Bitcoin is a digital currency whereas blockchain is the underlying technology behind Bitcoin.

Who invented Bitcoin?

There is no doubt about this. Satoshi Nakamoto, a secretive internet user or group invented bitcoin in 2008 before it went online in 2009. Many attempted efforts to find out who or what Satoshi really is have been proven futile.

What can Bitcoin be used for?

People see value in money that is out of the governments' and banks' control. Powered by the blockchain technology, Bitcoin has been seen as a tool for private, anonymous transactions and thus is the preferred choice of payment for shady activities like drug deals and other illegal purchases.

Is it worth anything?

Bitcoin is highly valued by the market. As of December 2016, there are around 16 million pieces of bitcoins in circulation. In March 2017, the value of Bitcoin, at $1,268, exceeded that of an ounce of gold ($1,233) for the first time. As of July 2017, the value of Bitcoin is still increasing. However, Bitcoin's value is actually determined by the value perceived by the public so it can be relatively volatile. There are people whom feel that Bitcoin is too much of a hype and foresees it to crash.

What is cryptocurrency?

Cryptocurrency is basically digital money that is designed to be secure and in many cases anonymous using the blockchain technology. It is available only over the internet and cannot be printed out as a physical object. It is also a currency associated with the internet that uses cryptography, the process of converting legible information into a code that is almost impossible to crack.

Cryptography was born out of the need for secure communication since the Second World War. It has evolved in the digital age with elements of mathematical theory and computer science to become a way to secure communications, information and money online.

The first cryptocurrency, needless to say is the well-known bitcoin and the second being Ethereum. Ripple and Litecoin are also the more famous cryptocurrencies in the market at this point of time. There has been a proliferation of cryptocurrencies in the past decade and there are now more than 900 available on the internet as of 11 July 2017.

How do cryptocurrencies work?

Cryptocurrencies are digital money that use decentralised technology to let users make secure payments and store money without the need to use their name or go through a bank. They run on a distributed public ledger, which records all transactions spent and held by currency holders.

Just like precious metals, cryptocurrencies are created through a process called mining, which involves using computer power to solve complicated maths problems that generate coins as a reward for the miners. Buyers can also purchase the currencies from brokers, then store and spend them using digital storages called cryptographic wallets.

Cryptocurrencies and applications of blockchain technology are starting to gain recognition in the financial sectors and more uses should be expected. The potential is actually quite huge and transactions including bonds, stocks and other financial assets could eventually be traded using this wonderful technology.

What are the most common cryptocurrencies?

There are more than 900 cryptocurrencies available over the internet as of 11 July 2017. This number is still growing and there is no sign of it slowing down. This is likely due to the success of Bitcoin which value has increased multiple folds since its introduction. The value of cryptocurrencies overall is at a record $137 billion as of 14 August 2017 according to Coinmarketcap.com. Bitcoin alone is more than $66 billion. By market capitalization, Bitcoin is without a doubt, the largest blockchain network, followed by Ethereum, Ripple then Litecoin. These cryptocurrencies are worthy of mention which we will discuss more as follows.

Bitcoin:

This cryptocurrency basically needs no introduction. It is the first and the most commonly traded cryptocurrency to date. The currency was developed by Satoshi Nakamoto in 2009, a mysterious figure or group who developed its blockchain.

Bitcoin has a market capitalisation of around $45 billion as of July 2017. By mid-August 2017, it is worth more than $66 billion. The value of a unit of Bitcoin as of 14 August 2017 has already crossed the $4,000 mark, definitely worth much more than physical solid gold. The price of Bitcoin is up more than 40% in August 2017, and more than 280% since beginning of 2017.

Bitcoin is designed to have only 21 million coins in total. There are currently about more than $16million of Bitcoins circulating in the market and it is estimated that by 2040, all 21 million Bitcoins will be mined. Bitcoins can be divided into smaller units, which the smallest is at one hundred millionth of a Bitcoin which is called a "Satoshi", named after the mysterious founder.

What happen after all the Bitcoins are mined?

Bitcoin can be celebrated by supporters or admonished by sceptics because of its finite supply. Supporters love Bitcoin's fixed supply because it likens back to the days of the sound money gold standard. Gold shares numerous similarities with Bitcoin, the most obvious being its fixed supply. Gold cannot be created out of nothing in simply any amounts. It must be extracted from the earth and put into circulation as market prices dictate. Bitcoin, if it ever achieves as widespread use as gold, can accomplish these same things with its own fixed supply.

Bitcoin takes gold's benefits a step further by being digital. The Bitcoin supply is not only incapable of being arbitrarily manipulated, it also eliminates the need for paper substitutes. With gold being heavy and taking up physical space, anyone in the right state of mind tends to prefer paper substitutes for gold rather than carrying actual bullion with themselves. This practice leaves gold stored in the bank, forcing people to trust the bank to handle their gold responsibly. As such, there is definitely a possibility that the gold stored in the bank might go missing as it is physical and still handled by humans. Bitcoin's digital nature eliminates this problem; since it costs almost nothing to store, and it takes up zero physical space in the real world. The best part now is that no one can touch it, not even you yourself.

Despite the promising benefits mentioned, there are some people whom still take issue with the fact that Bitcoin has a finite supply. One issue that sceptics like to harp on regarding Bitcoin's fixed supply is how miners will fare once they lose their block rewards. They worry that the mining system is unsustainable because once all the bitcoins are created, miners will have to rely on transaction fees to keep themselves financially operational. A reliance on miner fees instead of a block reward will make mining very unaffordable, which will lead to a contraction of miners, a centralization of the network, and possibly a complete collapse of the network.

Ethereum:

Developed in 2015, Ethereum is the second most popular and valuable cryptocurrency after Bitcoin. Ethereum has a market capitalisation of around $18 billion as of July 2017. However, Ethereum has had a turbulent journey. After a major hack in 2016, it split into two currencies. Its value has in recent months reached as high as $400 but crashed briefly to as low as 10 cents.

Like Bitcoin, Ethereum is a distributed public blockchain network. Although there are some significant technical differences between the two, the most important distinction to note is that Bitcoin and Ethereum differ substantially in purpose and capability. Bitcoin offers one particular application of blockchain technology, which is a peer to peer electronic cash system that enables online Bitcoin payments. While the bitcoin blockchain is used to track ownership of digital currency (bitcoins), the Ethereum blockchain focuses on running the programming code of any decentralized application.

In the Ethereum blockchain, instead of mining for bitcoin, miners have to work to earn Ether, a type of crypto token that fuels the network. Beyond a tradeable cryptocurrency, Ether is also used by application developers to pay for transaction fees and services on the Ethereum network.

How Ethereum Began

Four years after the introduction of Bitcoin, a 19-year-old by the name Vitalik Buterin thought of having a cryptocurrency based off Bitcoin. Vitalik is a programmer from Toronto and he first grew interested in bitcoin in 2011.

He co-founded the online news website Bitcoin Magazine in the same year, writing hundreds of articles on the cryptocurrency world. He went on to code for the privacy-minded Dark Wallet and the market place Egora.

Along the way, he came up with the idea of a platform that would go beyond the financial use of bitcoin. He released a white paper in 2013 describing an alternative platform designed for any type of decentralized application developers would want to build. The system is what we called Ethereum. Ethereum makes it easy to create smart contracts, self-enforcing code that developers can tap for a range of applications.

What is a smart contract?

Smart contract as the name suggests is a contract that self-executes, and the contract handles the enforcement, the management, performance, and payment. It is actually a computer code that can facilitate the exchange of money, content, property, shares, or anything of value. When running on the blockchain, a smart contract becomes like a self-operating computer program that automatically executes when specific conditions are met. Because smart contracts run on the blockchain, they run exactly as programmed without any possibility of censorship, downtime, fraud or third party interference.

Although all blockchains have the ability to process code, most are very much limited. Ethereum is different. Rather than giving a set of limited operations, Ethereum allows developers to create whatever operations they desire. This means that developers are able to build thousands of different applications that go way beyond anything we have seen before.

The Ethereum Virtual Machine

Before the creation of Ethereum, blockchain applications were designed to do a very limited set of operations. Bitcoin and the other cryptocurrencies, for example, were developed solely for the purpose of operating as peer-to-peer digital currencies.

Developers met with a problem. They can either expand the set of functions offered by Bitcoin and other types of applications, which is very complicated and time-consuming, or develop a new blockchain application and an entirely new platform. Recognizing this predicament, Ethereum's creator, Vitalik Buterin developed a new approach.

Ethereum has a core innovation, which is called the Ethereum Virtual Machine (EVM). Ethereum Virtual Machine is a "turing complete software" that runs on the Ethereum network. It allows anyone to run any program, regardless of the programming language as long as there is enough time and memory. The Ethereum Virtual Machine makes the process of creating blockchain applications much easier and more efficient than ever before. Instead of having to build an entirely original blockchain for each new application, Ethereum enables the development of potentially thousands of different applications all on one single platform.

Despite having a much lower value as of August 2017, Ethereum definitely looks more promising than Bitcoin based purely on the impact that it is expected to create. A future of unimagined possibilities lies ahead for Ethereum.

Ripple:

Ripple is another distributed ledger system that was founded in 2012. Unlike Bitcoin, Ripple can be used to track more kinds of transactions, not just cryptocurrencies. It has been used by banks like Santander and UBS and has a market capitalisation of around $6.3 billion.

According to the company OpenCoin which is behind Ripple, ripple addresses the need to keep money flowing freely. The purpose of Ripple, is to build on the decentralized digital currency approach which already has been laid out by bitcoin, and do for money what the internet did for all other forms of information

How would Ripple function like the internet?

Payment systems today are where email was in the early '80s. Every provider built their own system for their customers and if people used different systems they couldn't easily interact with each other effectively. Ripple is designed and aims to connect different payment systems together. The Ripple network is able to allow the seamless transfer of any form of currency, regardless whether if it is dollars, euros, pounds, yen or bitcoins.

Ripple, in many ways, is very much like Bitcoin. Similar to Bitcoin, Ripple's XRP unit is a digital form of currency based on mathematical formulae and has a limited number of units that can be mined. Both forms of currency can be transferred from account to account without the need for any redundant third party banks. And also, both provide the digital security to guard against the possibility of fake coins.

However, Ripple sees itself as a complement to Bitcoin rather than a competitor. In addition to giving Bitcoin more ways to connect with those using other forms of currency, Ripple promises expedited transactions and increased stability. As a distributed network, Ripple does not depend on a single company to manage and secure the transaction database. Consequently, there is no waiting on block confirmations, and transaction confirmations can go through the network quickly.

How many Ripples will there be?

OpenCoin targets to ultimately create 100 billion Ripples. Half of those are to be released for circulation, while the company plans to retain the other half.

Ripple doesn't receive transaction fees the way PayPal, banks and credit cards do. However, it does take a tiny portion of a Ripple which is actually equivalent to ~1/1000th of a cent from each transaction. The amount collected is destroyed rather than retained. The deduction is meant as to safeguard against the system being swamped by any individual who might try to put through millions of transactions all at once.

Litecoin:

To date, this currency is most similar in the form to bitcoin, but has moved much quickly to develop new innovations, including faster payments and processes to allow more transactions. The total value of all Litecoin is around $2.1 billion as of August 2017.

When Satoshi Nakamoto launched bitcoin as the world's first cryptocurrency, he made the code as open source, which means it can be modified by anyone and freely used for any other projects. Many cryptocurrencies in the market are pretty much modified versions of this code, with varying levels of success.

Litecoin was announced in 2011 with the goal of being the 'silver' to bitcoin's 'gold'. At the point of writing, Litecoin has the highest market cap of any mined cryptocurrency, after bitcoin.

Just like bitcoin, litecoin is generated through mining. Litecoin was created by a former Google engineer by the name Charles Lee. Charles saw some flaws in the Bitcoin and he was motivated to improve on that. Litecoin was therefore created. The main difference between Bitcoin and Litecoin is the block generation time. Bitcoin requires 10 minutes whereas Litecoin only requires 2.5 minute to generate a block. Since Litecoin can confirm transactions much faster than bitcoin, the implications are as follows:

- Litecoin can handle a much higher volume of transactions - thanks to its faster block generation. If bitcoin were to try to match this, it would require a significant amount of updates to the code that everyone on the bitcoin network is currently running. To put things in perspective, a merchant who waited for a minimum of two confirmations would only need to wait five minutes, whereas they would have to wait 10 minutes for just one confirmation with bitcoin. Simply said, Litecoin block generation is four times faster than Bitcoin's.

- The disadvantage of this higher volume of blocks is that the Litecoin blockchain will be proportionately larger than bitcoin's, with more orphaned blocks.

- The faster block time of Litecoin reduces the risk of double spending attacks – this is theoretical in the case of both networks having the same hashing power.

Transaction speed (or faster block time) and confirmation speed are often disregarded as a consideration factor by many involved in bitcoin, as most merchants can allow zero-confirmation transactions for most purchases. It is necessary to note that a transaction is immediate and it is just confirmed by the network as it propagates.

For miners and enthusiasts though, Litecoin actually holds a much more important difference to bitcoin, and that is its different proof of work algorithm. Bitcoin uses the SHA-256 hashing algorithm, which involves calculations that can be greatly accelerated in parallel processing. It is this characteristic that has given rise to the intense race in ASIC technology, and has caused an exponential increase in bitcoin's difficulty level.

Litecoin, on the other hand, uses the scrypt algorithm – originally named as s-crypt, but pronounced as 'script'. This algorithm incorporates the SHA-256 algorithm, but its calculations are much more serialised than those of SHA-256 in bitcoin. Scrypt favours large amounts of high-speed RAM, rather than raw processing power purely.

Why would you use a cryptocurrency?

Cryptocurrencies are known for being secure and providing a level of anonymity. Transactions in them cannot be faked or reversed and there tend to be significantly lower fees, making it more trusted than conventional currency. Their decentralised nature means they are open to everyone, whereas banks can be exclusive in who they will want to deal with.

As a new form of cash, the cryptocurrency markets have been known to take off meaning a small investment can become a large sum overnight. There has already been real life stories of millionaires created through investment of Bitcoin. But the same can work the other way. Therefore, people looking to invest in cryptocurrencies should be aware of the volatility of the market and the risks they take when buying. But relatively speaking, investing in cryptocurrencies seems like a much better bet than the lottery and stock market.

Chapter 4: Limitations of Blockchains

No human is perfect. Neither is technology. Blockchain though revolutionary, is not all glamorous and has issues of its own. Some people in the blockchain industry have pointed out that blockchain has become overhyped, when, in reality, the technology has its limitations and is not suitable for every type of digital interactions. Let us look at the main issues.

High Energy Consumption

Bitcoin and other cryptocurrencies consume a lot more energy than we had imagined. Cryptocurrencies blockchains are based on different proof algorithms to ensure a consensus in faceless transactions. For example, Bitcoin, the famous cryptocurrency that garnered the world's initial attention to blockchain technologies, uses a protocol called the proof-of-work system.

Although more efficient proof functions were developed since Bitcoin's creation, it's hard to reinstate new functions into programming running on older functions. As the blocks increases and the mathematical problems get more complex, computing energy consumption will inevitably increase significantly.

Human error

If a blockchain is used as a database, the information going into the database needs to be of high quality. The data stored on a blockchain is not inherently trustworthy, so events need to be recorded accurately in the first place. The phrase 'garbage in, garbage out' holds true even in a blockchain system of record, just as with a centralized database.

Unavoidable security flaw

There is one notable security flaw in bitcoin and other blockchains. There might be a possibility that more than half of the computers working as nodes to service the network gang together to make a lie which will make the lie become the truth. This is called a '51% attack' and was highlighted by Satoshi Nakamoto when he launched bitcoin. Due to this very reason, bitcoin mining pools are monitored closely by the community, ensuring no one gains such network influence.

No customer protection on the blockchain.

Blockchain technology operates as a push-based settlement system. This means that the individual holds power over the resource they want to verify on the blockchain. This could be anything from cryptocurrency to certificate authentication or land titles. The problem with this is if a transaction goes sour after it has already been verified on the blockchain, the only feasible way of returning the transaction is if all the parties agree to reverse it, which might require using a centralized system like a bank or lawyer. This will complicate matters and add extra cost to the transaction.

Settlement on a blockchain is slow.

Settling a transaction on the blockchain requires all the nodes in the network to come to an agreement that the transaction is valid. Transactions can be made instantaneously, however until the block in which the transaction is inserted in has been verified, it cannot be classified as trustworthy. In the time between a lodged transaction is made and when the block settles, a bad actor can launch fraudulent transactions to trick the network into what is known as a double-spend.

A very exciting upcoming technology that could solve this problem is the lightning network. This solution acts as a layer two of blockchain technology and can be applied to any public blockchain. It will enable instantly verified transactions for a fraction of the cost of today's settlement.

Selfish Miners

The mining process on the blockchain is an innovation which uses game theory economics to incentivise people to commit computer power for securing the network for a profit. The actual truth of this is generally miners won't care about settling as many transactions as possible. What they are interested in is how to make the most money by finding and verifying a block in the fastest way possible.

This leads to a problem of miners finding empty blocks and validating. There is also another problem known as Selfish Mining, which is a situation whereby a miner or mining pool finds and validates a block and does not publish and distribute a valid solution to the rest of the network.

The growing blockchain size

With every new block, a blockchain grows. This can be an issue because each node that is validating the network needs to store the entire history of the blockchain in order to be a participant. This is already a tough enough issue with the bitcoin blockchain where the transaction size is only a few bytes, the total blockchain size as of January 2017 is 98 gigabytes. Given that at the same time in 2016 the size was just 50 gigabytes, and the use of the blockchain is continuing to increase, the size can only get massive.

One of the biggest discussions in the bitcoin industry is whether the block size should be increased. If a blockchain has bigger blocks the blockchain size will definitely increase faster, therefore weeding out the small solo miners eventually. This is a big issue because the health of a blockchain network is partially dependent on the amount of nodes in the network, and the spread of those nodes across the world. The counter argument for this problem is that with sufficient advancement of technology, hard disk space will be very cheap in the future and will stay ahead of the blockchain size. The debate is going on and on without definitive conclusion. Only time will prove which school of thought is right.

Eventually settlement on the blockchain will not be cheap

Space in a block is a finite resource on any public blockchain. As the network utilization increases, the amount of transactions that will want to settle in a block will exceed the storage capacity. Public blockchain networks already have a solution built in for this, which is that transactions with a higher miner fee attached will get precedence to be included in a block. This makes total sense due to the fact that the miners want to maximize their profit so that they will include transactions with the highest fees first.

Do note that this is a feature, not a bug. Think about it. If settling on the blockchain was to be free, there would be far too many ways of attacking the blocks with dust transactions and blocking up the network. Originally the bitcoin blockchain did not have any block size limit but was eventually set to 1 megabyte to avoid a Sybil Attack on the network.

All of such problems have potential solutions that can be implemented as a fix. Some in the blockchain space feel that blockchains will eventually have layers of centralisation like the lightning network. This may not be a bad thing so long as there is a sufficient amount of encryption to protect the privacy of the people who want to use the centralized layers of the network.

All Bitcoins Have Mined

It is not wrong to think that once all the bitcoins have been mined, transaction fees will be the only sole source of income for miners. The main concern, then, is whether or not transaction fees will be enough to keep miners financially afloat.

Using current mining costs as a measure of required mining profitability over 100 years from now is definitely unreliable, since we cannot be certain how mining technology will progress over time. It is entirely possible that mining chips will become so small and cheap that they can be installed on all electronic devices. This development would turn mining from a purposeful business decision to an afterthought, surviving in the background of daily life. Furthermore, mining hardware may become so energy efficient over the next century that transaction fees prove to be more than enough to keep miners in business.

Looking from another perspective, it may also be possible that transaction fees simply rise to a level sufficient for mining profitability. If, once all the bitcoins have been mined and the entire world uses the digital currency as its primary medium of exchange, then it is possible that transaction fees will rise due to an increase in the demand for transactions.

However, the likelihood of fees rising to such a rate is very unpredictable at this point, since the consensus in the community at present is to have a gradually increasing block size to ensure network scalability. This means that, if the block size continues to grow, people will always be able to have their transactions confirmed at low fees. This prospect may seem like a threat to the network on the surface, as it entails forcing miners to survive on low fees after the block reward is gone. But not increasing the block size may be an even larger threat to the network than low transaction fees. If blocks reach their maximum size, no more transactions can be confirmed until a new block is created, which means excess transactions will be dropped from the network. This scenario may mean having higher fees for miners since people can only agree to pay higher fees in order to get their payments through. But it would also greatly discourage people from using Bitcoin altogether, which could kill the digital currency much faster than a centralized mining network.

Although Bitcoin's fixed supply means that miners will eventually have to give up their block rewards, it also creates an opportunity for miners to survive on transaction fees through simple monetary theory. Once all 21 million bitcoins have been mined, the supply cannot increase regardless of growing demand. The result of this discrepancy between the supply of and demand for money is a steady and gradual decrease in the general price level, which equates to an equally steady and gradual increase in the purchasing power of money. Therefore, as bitcoin miners collect transaction fees over time, no matter how large or minute, the funds gain value. This value appreciation across time turns fee-centric mining into a financially infeasible task to a sensible, long-term investment.

In conclusion, the problems or issues with blockchain mainly lie on its sustainability. The current success of blockchain concept cannot be denied as we can take reference from how much Bitcoin is worth now in 2017. However, the future is still uncertain as it is anyone's guess if the blockchain concept can remain sustainable once it turns ridiculously massive.

Chapter 5: Applications

Blockchain technology holds great potential to transform business operating models in the long term. There are people who might disagree but blockchain distributed ledger technology is more of a foundational technology than a disruptive technology. Disruptive technology typically attacks a traditional business model with a lower-cost solution and overtakes incumbent firms quickly. Blockchain technology does not work in this manner but rather, promises to bring significant efficiencies to global supply chains, financial transactions, asset ledgers and decentralized social networking.

Blockchains technology can be integrated into many different areas. As blockchains can be thought of as an automatically notarised ledger, they alleviate the need for a trust service provider and are predicted to result in less capital being tied up in disputes. Blockchains have the potential to reduce systemic risk and financial fraud. They automate processes that were previously time-consuming and done manually.

The most well-known application for blockchain now is basically in cryptocurrencies. However, blockchain applications can be vast. Banks are interested in this technology because it has the potential to speed up back office settlement systems. For example, UBS are opening new research labs dedicated to blockchain technology in order to explore how blockchain can be used in financial services to increase efficiency and reduce costs.

The Big Four

Each of the Big Four accounting firms is testing blockchain technologies in various formats. Ernst and Young has provided digital wallets to all of the Swiss employees, installed a bitcoin ATM in their office in Switzerland, and accepts bitcoin as payment for all its consulting services. However, the other threes like PwC, Deloitte, and KPMG have taken a different approach from Ernst & Young and are all testing private blockchains.

The big companies have taken the lead to study on how blockchain can help with their business. Let's look at the other industries or places that blockchain application can be used.

Smart Contracts in the Legal Profession

There is a digital revolution going on in the legal industry and blockchain is the technology leading this transformation. The law is being digitized. If you have ever had to close a mortgage or been part of any legal dispute you know that lawyers are good at creating tons of paperwork.

If we can digitize the process of keeping track of the paper trail then it will reduce the cost and potential for human error. It could change how things work forever. Imagine recording everything on a shared ledger that becomes irrefutable digital proof that this legal event happened between two parties.

Legal issues might not strictly apply to mortgage only. It could be anything from a marriage to a divorce proceeding; a property transaction to a land reclamation; and anything else that involves digital proof. Cutting costs out of the legal system from administration to time would be a game changer for the legal profession.

Clearing and Settlement

Now this is probably one of the most interesting potential uses for blockchain when it comes to finance. Clearing and settlement costs the financial industry billions of dollars yearly.

It is estimated that blockchain technology could save the industry $20 billion a year which would be a huge benefit to consumers in lower fees for transacting things like wire transfers, overseas purchase and clearing fees on investment trades.

To make this even clearer, let's look at a simple example. Currently if you sell a stock, the transaction takes 3 days to complete and settle for the money to show up in your account. If we see a universal adoption of blockchain technology in the financial market, that transaction settlement and clearing could go from 3 days to an almost immediate process.

Blockchain is destined to become the backbone of how all financial transactions occur at some time in the not too distant future. It will require a major technology upgrade for the entire industry, but the benefits and savings of time and money will be too huge to ignore.

Supply Chain

Imagine items bought and sold in the supply chains of the world being recorded in near real-time on a shared ledger. You can record virtually any other information you want like destination, who is shipping it, when it reaches port, tax and anything related to the shipment. That is why it's a smarter tracking system. The companies that invest and adopt this sort of system will have a competitive advantage in the future.

Voting

Applying blockchain to voting is never a new topic. Politics can be a dirty game and it is highly possible that the vote counts are vulnerable to human errors and cheating. Blockchains can serve as the medium for casting, tracking, and counting votes so that there is never a question of voter-fraud, lost records, or fowl-play. By casting votes as transactions within the blockchain, voters can agree on the final count because they can count the votes themselves, and because of the blockchain audit trail, they can verify that no votes were changed or removed, and no illegitimate votes were added.

Healthcare

Healthcare institutions suffer from an inability to securely share data across platforms. Blockchain can allow hospitals, payers, and other parties in the healthcare value-chain to share access to their networks without compromising data security and integrity.

Insurance

It is not unusual for insurance to succumb to frauds and illegitimate claims with the current system. However, smart contracts can provide transparency and an irrefutable record for managing claims if it was put on a blockchain and verified by the network. This would basically mean that identical claims would not be paid twice or more thus reducing the potential for fraud and duplicate claims being paid. Blockchain would allow for a more streamlined claims process that would be beneficial to the customer experience as well as save money for the insurance company.

There are many more areas that we can apply blockchain to. The above are just some small examples of the potential uses for blockchain. As the world continues to evolve, the potential of using blockchains in other areas will definitely emerge and also increase.

Conclusion

The blockchain has been a hot topic for discussion as it has the potential to revolutionise domestic and international transactions. It is most well known for being a public ledger of all the Bitcoin transactions, which continues to grow exponentially. Blockchain allows individual parties to transact securely in the absence of a third party intermediary and it is clear that some businesses recognise the potential savings if they can just remove the third party intermediary.

The blockchain technology removes the need for "trust". Thousands of computers around the world are each holding a copy of the blockchain history record. There is no official copy and no computer is seen as more valid than another - they each mutually verify the ledger and there is no centralised authority such as a government or a bank. This decentralisation is one of the revolutionary aspects of the technology as it distributes power to everyone instead of just to an individual.

"Mining nodes" are computers connection to the blockchain; they race to validate transactions, create new blocks and have these accepted by the network. Within the context of Bitcoins, the successful computer is rewarded in bitcoins and generally there is a reward for miners to work with each different cryptocurrencies. Once accepted, each new block is sealed permanently and contains a link to a chain of prior blocks, making the chain more secure. Blocks can be added to the ledger but cannot be removed or corrupted.

The enhanced security of distributed ledger technology can benefit many industries including banks and financial institutions. For example, the system can facilitate the effective and secure transfer of ownership of digital assets (such as shares and bonds). The open ledger nature of the blockchain means that financial institutions may be able to monitor a customer's payment history more accurately, in the context of making a decision to lend. From a customer's perspective, the blockchain represents a more transparent system with the internal mechanisms not subject to "behind closed door" processes.

Not all is rosy for the blockchain applications. There has been inevitable criticism about the public nature of the blockchain and its encroachment on individual privacy. Blockchain technology has also been criticised as a forum for use in illegal activities, and in particular the transfer of the proceeds for crime. There has been some questions as to the potential applications of such technology beyond bitcoin mining. Its distributed and decentralised nature makes it inherently more cumbersome than some other softwares especially as the "chains" grow in size. This will definitely take up humongous memory space as it keeps growing. Energy consumption will also significantly increase as the need for faster and more powerful computing power increases. Organisations will have to give considerable thought as to where and how to apply the blockchain before jumping head down into it. Nevertheless, the technology and/or its derivatives represent a real opportunity to all those connected with the future of business.

As we all know, blockchain is the underlying technology behind cryptocurrencies like Bitcoin and Ethereum. Over the years, cryptocurrencies have gained much attention from the world and some of them like Bitcoin especially, are worth a lot of money now. However, the real focus should never be on Bitcoin but on the underlying technology itself instead. Blockchain technology is not limited to just being used in cryptocurrencies, but to any digital data that can be stored, distributed or transacted like property titles, music, insurance, physical goods and assets, even your digital identity. Researchers and investors are talking about how blockchain technology will be the next big thing across industries like finance, retail and even healthcare.

Blockchain technology can fundamentally change how we exchange value and perhaps that's why this has caught everyone's attention. This is still in its nascent stages but definitely a technology that holds vast promise and something to watch for, in the near future.

Author's Notes

Thank you again for downloading this book!

I hope this book was able to help you to understand blockchain and Bitcoin better.

Finally, if you enjoyed this book, then I'd like to ask you for a favour to leave an honest review for this book on Amazon so that other readers can decide if this is the book that will be applicable for them. Very much appreciated!

Thank you, good luck and all the best!

Best Regards,

Shane Bock

www.ingramcontent.com/pod-product-compliance
Lightning Source LLC
Chambersburg PA
CBHW050019230526
45470CB00003B/1042